Lester made a hit.
"Do you want to try,
Fern?" he asked.
Fern was thrilled.
"Yes," she said.

1

"I'm your big brother,"
said Lester. "I can do lots of
things you can't. But I can
help you."

"Here's my other top," said Lester. "It has numbers and letters on it. But it is not the right size."

Lester gave her the light
bat. "Put your right hand here,"
he said. "Put your left hand
under it."

"I'm going to toss it,"
said Lester. "Swing the bat
after that."
Fern wanted to try.

"You might miss it," he said. "But I'll just toss another one to you. You'll get better and better."

Lester tossed the ball.
Bam! It went high, high,
high!

"Good job!" said their mother. "Only a big brother could make that happen."

Lester gave Fern a high five.

The End